GUARDIANS OF
AIR FORCE RESOURCES

Guardians of Air Force Resources
Copyright © 2023 by Ray Acuña

Published in the United States of America
ISBN   Paperback: 978-1-960629-60-9
ISBN   Hardback: 978-1-960629-65-4
ISBN   eBook: 978-1-960629-61-6

All rights reserved. No part of this publication may be reproduced, stored in a retrieval system or transmitted in any way by any means, electronic, mechanical, photocopy, recording or otherwise without the prior permission of the author except as provided by USA copyright law.

The opinions expressed by the author are not necessarily those of ReadersMagnet, LLC.

ReadersMagnet, LLC
10620 Treena Street, Suite 230 | San Diego, California, 92131 USA
1.619. 354. 2643 | www.readersmagnet.com

Book design copyright © 2023 by ReadersMagnet, LLC. All rights reserved.

Cover design by Ericka Obando
Interior design by Don De Guzman

Ray Acuña

Air Force Office of
Special Investigations Special Agent (Ret.)

Guardians of
Air Force Resources

For Roselia, my beloved wife of forty-five years, and for Elizabeth, Leeann, and Roxann, my three amazing daughters.

CONTENTS

Preface 1

Chapter One 3
    Mysterious Drug

Chapter Two 6
    Bogus Travel Claims

Chapter Three 9
    The Buy American Act

Chapter Four 12
    An Airman's Revenge

Chapter Five 15
    A Heist at the BX

Chapter Six 19
    Door-to-Door Marijuana

Chapter Seven 22
    The Masked Assailant

Chapter Eight 26
    Dirt-Bike Trails

Chapter Nine 29
    Jet Engine for Sale

Chapter Ten 32
    Air Force Beef Uninspected

Chapter Eleven 34
    Gardening Marijuana

Chapter Twelve 37
    Death of Three Soldiers

Chapter Thirteen..40
  IMPERSONATION AND MURDER
Chapter Fourteen 43
  THEFT OF MILITARY PROPERTY
Chapter Fifteen 47
  PAYMENTS TO A GHOST
Chapter Sixteen...49
  KOREAN MARRIAGE SCAMS
Chapter Seventeen...52
  CRITICAL DRUG INVESTIGATIONS

Memorable Moments/Awards 56
My Amusing Tidbits 65
Life After OSI 69
About the Book..73
About the Author 74

PREFACE

IN 1973, I became a special agent of the Air Force Office of Special Investigations (AFOSI), and I retained that assignment until my Air Force retirement in 1987. While in that office, I also became an integral member of a team of select few special agents who were assigned to many activities worldwide at Air Force installations. OSI agents stand with confidence, ready, willing, and able to meet any challenges presented by criminal activity, fraud matters, counterintelligence assignments, and dignitary protection operations.

When OSI agents are dealing with all these challenges, they are essentially "Guardians of Air Force Resources." In this memoir, I will describe a variety of investigations in which I was involved and a typical information report that OSI agents are provided daily worldwide. Each chapter takes the reader through the meticulous investigative steps necessary to satisfy the elements of proof.

I have also included a section of some of my most memorable moments and awards. The last OSI section, which I call "My Amusing Tidbits," is remarkably interesting. Lastly, I detail what profession I selected after the OSI and the strong correlation between the two careers.

CHAPTER ONE

MYSTERIOUS DRUG

THE YEAR WAS 1973. One summer day on an Air Force installation in the state of Texas, I received a call from a confidential source of information (CSI) who asked that I meet with him. When I did, he began to give me information about an airman who had recently returned from the country of Turkey. The airman, who resided in the barracks, was selling samples of a Turkish substance for ten dollars per gram. The CSI said the buyers of the brown, powdery substance would smoke it in pipes and seemed to experience a "high" and hallucinate. I asked the CSI if he could find out where the airman was holding the substance.

The CSI later reported that the airman would go to his car, presumably owned by the airman, and return with a small amount of the substance. In each case the CSI witnessed, the sequence was the same. I asked the CSI to further identify the car the airman would go to in order to retrieve the brown, powdery substance.

The CSI later came back with the description of the vehicle: a blue 1965 Ford Mustang, license plate (XXX). A search of motor vehicle registration records revealed that the Ford Mustang did indeed belong to the airman in question.

The base commander was briefed on the current details of the investigation. He requested further investigation into the airman in question. The staff judge advocate (SJA) was also briefed, and I asked the SJA if there were sufficient grounds to obtain an "authority to search and seize" (a warrant) in order to fully conduct a search of the 1965 blue Ford Mustang belonging to the airman in question.

The SJA advised there were. After coordinating with the base commander, who is responsible for all personnel and property on a military installation, a signed authority to search and seize was issued to search the 1965 blue Ford Mustang.

Another agent and I executed the warrant in the presence of the airman in question, the owner of the vehicle. The inside of the vehicle appeared to be clean, except for a four-inch briefcase, which was locked. The airman was asked to open it and reluctantly complied. Inside the briefcase was a twelve-inch, sealed-top plastic bag containing a brown, powdery substance. I noticed a cinnamon scent when the bag was opened. The briefcase was seized as "evidence under proper receipt." The agents continued to search under the front and rear seats for any possible concealment openings.

I looked one last time around the middle console. In front of the gearshift, there appeared to be a six-inch-by-six-inch leather cover attached to the dashboard. After closer examination of the cover, it appeared to be removable. Removing the cover revealed a small compartment in which several homemade, ceramic smoking pipes were found. One pipe had the shape of a king's head, crown included. All pipes had the scent of the brown, powdery substance and were seized as "evidence under proper receipt."

The powdery substance was tested using the BD (Becton, Dickinson and Company) field test used by OSI and other law enforcement agencies. There was a reaction, but the test for marijuana and crack cocaine was inconclusive. The substance was taken to the local Drug Enforcement Administration (DEA) office. Coincidentally, one of the DEA agents had been assigned in the vicinity of Turkey and recognized the brown, powdery substance. He described it as hashish—the flowering tops and leaves of Indian hemp that was smoked, chewed, or drunk as a narcotic and intoxicant. When the dried leaves of the hemp are ground into a brown, powdery substance, it yields hashish in its purest form. The DEA agent said that hashish was very much available in the streets of Turkey. Currently, the DEA classifies it as a schedule 1 substance which is illegal to use and possess.

The brown, powdery substance seized from the vehicle of the airman in question weighed eight grams less than a pound. The airman was selling the drug at ten dollars a gram. The seizure cost him around $5,000. I believe this amount of hashish was the largest seizure of an illegal drug on a military installation in 1973. The airman in question was subsequently court-martialed and charged with use and possession with intent to distribute. He was dishonorably discharged and received "bad time" at Fort Leavenworth, Kansas.

CHAPTER TWO

BOGUS TRAVEL CLAIMS

IN THIS CHAPTER, I explain the compelling need to have confidential sources of information (CSIs) assigned to important Air Force resources—for example, contracting, logistics, base exchanges, base commissaries, accounting and finance offices, aircraft maintenance, automation offices, and others. In the accounting and finance office, I happened to have an excellent CSI who strongly believed that "honesty was the best policy" when it came to filing and claiming submitted travel vouchers.

The year was 1974. One day, the CSI called me and reported that after reviewing several travel vouchers, he believed the traveler's lodging claims at the temporary duty (TDY) airbase "smelled a little fishy." I later met with the CSI, and he explained why there was a question on the travel voucher claims. The CSI said that the lodging receipts submitted by six firemen were possibly written by the same person in the amount of $200 each. The receipts reflected the name of mobile home rentals located off base. The six firemen were undergoing certification training at the TDY base for two weeks. These claims were authorized. However, when the CSI contacted the mobile home rental business to inquire about the number of bedrooms in the mobile homes, he was told they had three to four bedrooms. The CSI was told it was odd that each fireman rented such a large mobile home.

The following day, I went to the accounting and finance office and obtained the six firemen's travel vouchers under proper receipt. After subsequent review of the documentation, I went to brief the base commander, and he requested further investigation. One of the flexibilities of OSI is when investigations cross state lines or foreign countries' borders, leads can be forwarded to the nearest OSI detachment for assistance in the original investigation. Since in this case there was an OSI detachment at the airbase where the firemen were receiving training, I forwarded my report to them and asked that they interview the person who signed the $200 receipts to ascertain the validity of them.

As a result of my CSI's lead, the individual who originally signed the lodging receipts was interviewed. During the interview, he admitted the receipts were bogus. The interviewing OSI agents immediately advised the individual of his rights. In a written statement, the individual explained how he had rented the six firemen two mobile homes and had charged each fireman $25 per week. The amount for lodging totaled $300 for the two-week training period. The individual claimed the firemen talked him into providing them the bogus $200 lodging receipts. In return, the firemen promised him an additional $100 when they returned to their home base and settled their claims for reimbursement.

In return for his full cooperation and written testimony concerning the six firemen, the local FBI office elected not to file charges against the individual who signed the bogus receipts. This information was passed on to the owner of the mobile home rental park; he fired the individual involved.

After I received the written testimony, I began interviewing the six firemen who were civil service employees. I started with the oldest member of the group, a military retiree. After advisement of rights, he was willing to admit the use of the bogus lodging receipts to claim reimbursement from the government. He said he was quite sure the OSI had talked to the person who made out the bogus receipts because he had noticed the written statement next to me when I was interviewing him. He acknowledged being familiar with the statement in the DD Form 1351-2 Travel Voucher that reads, "There are criminal and civil penalties for knowingly submitting a false, fictitious, or fraudulent claim, U.S. Code Title 18, Section 287 and 1001."

The other five firemen, after advisement of rights, were subsequently interviewed. Each provided written statements giving essentially the same account as the oldest member. The five firemen were also aware of the statement above the signature block of the voucher form.

All six firemen appeared before the US magistrate at the same time. They were found guilty as charged, and each received a sentence of six months plus a $500 fine. The sentences were suspended, and the firemen were placed on unsupervised probation for five years.

CHAPTER THREE

THE BUY AMERICAN ACT

THE YEAR WAS 1975. I met with a confidential source of information (CSI) who reported that at the air base there was a contractor responsible for constructing a large building which would house the undergraduate pilot training flight simulator. The source continued to say he was told that the aluminum steel the contractor was using in the construction of the building was more inferior then American-made aluminum steel. The aluminum steel the contractor was using had markings that indicated it had been made in France.

Additionally, the contractor had hired skilled craftsmen from across the border in Mexico. Using authorized work permits, the craftsmen were properly crossing into the United States. However, the CSI was told by some of the craftsmen from Mexico that the contractor was paying them half of the minimum wage authorized by law. The contractor was paying the Mexican skilled craftsmen only five dollars an hour, instead of the ten dollars an hour required by the Department of Labor (DOL) at that time.

With information I received from the source, I went to the contracting office and spoke to the chief of contracting, who had assisted me with contract files in the past and whom I knew to be trustworthy. I immediately wanted to know if the contractor was subject to the Buy American Act when purchasing materials to be used for the construction of the building. Yes, he was. Additionally, the contractor was to pay all employees at least the minimum wage established by the US Department of Labor.

After I reviewed the contract, the base commander was briefed on the details of this matter, and he requested an investigation. Because the subject of this investigation was nongovernment, a civilian, it was time to get the federal authorities involved. I contacted the nearest office of the DOL and was told that Special Agent Marion *Jones (last name changed)* would come to my office.

The next day, I was in my office when my secretary came in and said Special Agent Jones was there to see me. I said, "Tell her to come in." The secretary had a big grin on her face.

In the doorway to my office came a man in western attire; he was about six foot five and weighed about 250 pounds. He said, "I'm called Buck. I am from the DOL."

I thought to myself, *I will never tell Buck I thought he was a female agent.*

I briefed Agent Jones on the details of this investigation. Agent Jones said he had the authority to examine subjects' pay records. I took Agent Jones to the subject's construction site. The subject was not there. However, the subject's foreman was in one of the construction trailers on the site. Agent Jones spoke with the subject's foreman and requested the payroll records.

Agent Jones said that after a review of the subject's payroll records, he believed there was evidence of irregularities in the hourly wages paid skilled craftsmen from the US versus those from Mexico. He said both US and Mexican craftsmen should receive the same hourly wages.

Agent Jones seized the subject's payroll records as evidence and subsequently presented the case to a US magistrate. The judge found the subject guilty of violating DOL laws. The contractor was ordered to make restitution to his employees in the amount of $75,000 and to pay the court a hefty fine. He was placed on probation for several months. As for the violation of the Buy American Act, the air base would have to address that issue. In the meantime, the contracting office suspended the subject's construction project pending the DOL investigation.

Once the subject made restitution to his employees in the amount of $75,000, he was instructed by base officials to remove all the aluminum steel he had installed in the project. It was at that time that the air base contracting office terminated the contract worth almost one million dollars. I imagine the subject was not a happy camper after losing the contract.

Soon after he lost the contract, I received a call from the subject; he wanted to meet me on an isolated road. I told him I had nothing to say to him. Besides, according to contracting personnel, the subject was described as wearing a floor-length mink coat and a funny hat, and driving a big, pink Cadillac.

CHAPTER FOUR

AN AIRMAN'S REVENGE

THE YEAR WAS 1975. Early one morning, I got a call from the Security Forces desk officer. He said that I may want to respond to the flight-line parking lots for some vandalism that had taken place there. It was seven in the morning, just before daybreak.

When I arrived at the scene, there were several police cars with their emergency lights flashing. The police cars were blocking the entrances to the parking lots. Also, crime-scene tape was strung up around the parking lots to preserve any evidence. Security forces are trained to respond to incidents, and if there are no lives to attend to, then the incident site is protected and preserved until the arrival of the investigators.

Okay, what was the vandalism in this case? I was almost completely floored when I stood in front of about thirty-nine privately owned vehicles in the parking lot; their windshields had all been smashed to pieces. I thought to myself, *Where do I begin to process the crime scene?* I knew all the owners of the vehicles were standing by to see when the vehicles could be released so they could call for windshield replacements.

Examination of all the broken windshields disclosed one possible similarity, and that was the object used to cause the breakage had to be larger than a carpenter's claw hammer or a mechanic's ball peen hammer. Windshields are made of several layers of glass, which gives them some resistance. A claw hammer or ball peen hammer could have done the damage, but it would require many blows to effect the damage done to those windshields. The damage in this case was most likely done by a much larger hammer, such as a sledgehammer.

After I completed processing the vehicles, I found no traces of any evidentiary value. I released the vehicles to their rightful owners. The estimated cost of damage in this case was based on the price to replace a windshield at that time, around $200. So, for the thirty-nine damaged cars, it came to a total of $7,800.

The base commander was briefed on this incident, and he requested an investigation. In the days that followed, I attempted to identify anyone who might have seen a person or might have heard any noise in the vicinity of the flight-line parking areas. All efforts to find a witness were unsuccessful, but I was never one to quit on any investigation.

One day, the base mental health officer stopped by our office, as she had done on previous occasions. She had expressed an interest in joining the OSI, and we were having ongoing conversations about it. On that occasion, I asked her what type of a person would possess an angry disposition intense enough to destroy the windshields of thirty-nine vehicles. The mental health officer indicated that possibly she was counseling such an airman; however, she could not reveal who he was because of the doctor-patient relationship. I acknowledged that fact, but asked if she could tell me if the airman belonged to an organization on the flight line. She nodded her head yes and departed.

The flight line has three organizations and over 300 military members. I knew the first sergeants of those organizations. One of them was a close friend of mine. I talked to him and asked him if he had any airman with disciplinary issues. He had an airman who was disgruntled with his supervisors and disliked authoritative figures in the military. The first sergeant gladly provided the name of the airman. I asked the first sergeant if he could have the airman come by my office. The airman did so.

Before I began to interview the airman, I advised him of his rights and told him that he was suspected of causing damage to vehicles in parking lots. The airman did not request a lawyer and denied any knowledge of the damage to vehicles. I then asked the airman for consent to search his barrack room, which he gave in writing.

The first thing I spotted in the airman's barrack room was a large sledgehammer leaning next to a wall locker; its handle was thirty-two inches long, and it had a lot of scratches next to the metal part. I immediately put on some plastic gloves and seized the sledgehammer as evidence under proper receipt. I asked the airman whom the sledgehammer belonged to, and he said he had borrowed it from the base engineers but had brought it to his room before taking it to his duty section. The airman asked why I was taking the sledgehammer, and I told him that there could be glass fragments on the handle that might match glass taken as evidence from the vehicles' broken windshields.

Before I departed the airman's room, he admitted to smashing the windshields. I reminded the airman that he was still under advisement of his rights, but he did not want a lawyer. He was willing to provide me a written statement. The airman was court-martialed, received a bad-conduct discharge, and had to make restitution for the damages.

CHAPTER FIVE

A HEIST AT THE BX

THE YEAR WAS 1976. It was early Monday morning when the base exchange (BX) manager, *Jim (name changed),* notified the Security Forces desk officer that the BX had been broken into and that several jewelry showcases had been looted. Some police patrols responded. I was notified and responded as well. When I arrived on the scene, security forces were stationed at all entrances to the BX building to prevent anyone from entering.

I first interviewed Jim, who said he was the only one who opened the BX. He walked in and immediately noticed there were store displays disturbed and broken doors in the jewelry counters. Jim said that he had not touched or moved anything. He also pointed to the ceiling, where there was a grill hanging from the air vent, as a possible way that the suspect(s) might have entered the store. I asked Jim for a description of the jewelry that was taken.

Jim said, when his employees showed up for work, he would tell all of them, except the one who worked behind the jewelry counter, to come back at about 2:00 p.m. I reminded him that I would process the jewelry counter for latent fingerprints first, so upon arrival the clerk could start her inventory. Around the counter were a few empty boxes that would normally contain men's and ladies' wristwatches and other assorted jewelry pieces. I decided to take the boxes back to my office for processing for latent fingerprints. There were also empty tennis-shoe boxes on the floor. I prepared to take them back to the office as well.

I contacted the base photographer who had worked in a small-town police department as a crime-scene photographer. He was especially useful in many cases. He arrived, and we finished inside the store and then went on the roof. There was a base civil engineering utility truck next door, so I borrowed their ladder. The photographer and I climbed the ladder to the roof of the store. I walked straight to the large air vent and immediately noticed its panel door lying down on the roof next to the vent.

I noticed what appeared to be several tennis-shoe footprints on the door. I always carried different sizes of print-lifting tape, so I used two sheets of it to lift the shoe prints. The panel door was somewhat dusty, so the shoe prints showed up perfectly clear. The photographer took pictures of the shoe prints in a one-to-one scale. When I turned the air vent door over while wearing gloves, I spotted latent fingerprints, which I also lifted. As I looked inside the air vent, I could see the inside of the store and the grill that was pushed or kicked open by the alleged suspect(s). More photographs were taken, and we were done on the roof.

The next day, Jim came by my office and reported what had been taken. Missing from the jewelry counter were men's wristwatches and gold rings, ladies' wristwatches and gold rings, bracelets, and necklaces. Also, several tennis shoes were stolen. The total cost came to $10,000.

A few days passed, and I still had several latent footprints and fingerprints, but no suspect(s) to match them to. In the base newspaper, I placed ads describing what was stolen and requesting I be contacted at the OSI if anyone had seen anything or had any information about the theft.

A couple of days later, I received a call from a mother in base housing. She had discovered some jewelry (bracelets and necklaces) in her fifteen-year-old daughter's room. I asked her to come to my office with the jewelry, her daughter, and her husband. They did so that same afternoon. I reminded the parents I could not ask their daughter any questions without their consent, and they both gave it.

Using the list I got from Jim, I could readily see the jewelry brought into my office, particularly the necklaces, fit the descriptions of some of the items on Jim's list. I then asked the young daughter where she got the jewelry.

She was a little timid and seemed afraid to answer, but her mother jumped in and said, "Tell him where you got the jewelry."

The girl replied, "From my boyfriend."

I learned her boyfriend's name; he was fifteen years old. The girl said her boyfriend and his friend, also fifteen years old, would meet a Sergeant *Joe (named changed)* at the bowling lanes on base to talk about breaking into the BX. The sergeant was later identified as Staff Sergeant (E-5) Joe, Civil Engineering Squadron. Staff Sergeant Joe was a heating and air conditioning mechanic. The young girl's parents were asked if the daughter could provide a written statement; they both agreed.

The parents of both boys were instructed to bring their sons to see me at the OSI. Again, I told the parents I needed their permission to question their sons. They granted it. I asked the boyfriend what he knew about the break-in at the BX. As he was staring at the jewelry taken from his girlfriend, he recounted that he and his friend both entered the BX through the air vent on the roof. I asked what Staff Sergeant Joe's part was in this breaking and entering. The boyfriend said Staff Sergeant Joe planned the whole thing. He said Staff Sergeant Joe also provided a sketch that showed the location of the air vent; he told them it would be unlocked.

"Where is the sketch, and what was Staff Sergeant Joe getting out of this deal?" I asked.

He said that the sketch was at his house and that Staff Sergeant Joe was going to get half of the jewelry. However, the boyfriend had not given any of the jewelry to Staff Sergeant Joe after meeting with his girlfriend and giving her a few pieces. On his walk back home, he carried the rest of the jewelry in a felt bag. He walked by a barrack and threw the bag on top of it. I asked his parents' consent to get a written statement from their son. They gave it. In the event I could lift the latent fingerprints of Staff Sergeant Joe, I was anxious to get the sketch he had made. I went with the boyfriend and his parents to retrieve it.

I then interviewed the friend in the presence of his parents. He essentially provided the same information. I got a written statement from him with his parents' consent. The next thing I did was go after the bag of jewelry on top of the barrack identified by the boyfriend.

At the barrack the boyfriend identified, I looked on the second floor for the utility room that had a trap door with a ladder to the roof. However, the trap door was locked from the roof side. This made no sense. I contacted the fire department to borrow an extending ladder to reach the second floor of the barracks. I started to climb as a couple of firemen stood below. As I got to the extended part of the ladder, it began to loosen from the hooks that lock when the ladder extends. I held on tight. It slipped down only about four feet, but the feeling was scary. One fireman climbed up and asked if I was okay. I said, "Yes, let's try it again, but let's make sure the hooks are secured!" When I reached the roof, I walked a few feet and spotted the bag of jewelry.

I finally brought in Staff Sergeant Joe for an interview. He was advised of his rights, and he requested a lawyer. Staff Sergeant Joe's lawyer from the Area Defense Counsel (ADC) requested to see all the evidence against staff sergeant. The ADC lawyer coordinated with the staff judge advocate (SJA) and made a deal in exchange for a confession. Staff Sergeant Joe would not get a bad-conduct discharge; however, he was reduced to the lowest rank, airman basic, and ordered to pay damages to the BX. The parents of the boys and girl involved were ordered to move off base.

CHAPTER SIX

DOOR-TO-DOOR MARIJUANA

THE YEAR WAS 1975. I received a call from a confidential source of information (CSI) to meet with him. I met with the CSI, and he related that a female airman named *Susan (name changed)* had offered to sell him marijuana. Susan was further identified as a twenty-eight-year-old medic who worked at the base hospital. She was a few years older than the average female enlisting in the Air Force at the time. The CSI said Susan bragged about being busted by the local police in an Eastern state for possession of marijuana, but the judge dropped the charges.

The CSI was instructed to make the "buy" of marijuana from Susan while the OSI surveilled the transaction. The OSI called this a "controlled buy." At the time, marijuana was being sold in "baggies" (plastic bags used for sandwiches). The going rate for a baggy was $20.

The day came when Susan had marijuana to sell. The CSI agreed to meet Susan in the parking lot of the hospital barracks at a certain time of night. The CSI was provided money, and the controlled buy took place under observation by me and another agent. Afterwards, we met the CSI and took the baggy of marijuana as evidence. The CSI said Susan was carrying a blue cosmetic case from which she removed the baggy of marijuana. The CSI said Susan would have more marijuana later. I was looking forward to making a second controlled buy.

In the meantime, I met with another CSI from a different organization who advised me that he had run into Susan in the male barracks. The CSI said females could visit the male barracks at certain times of the day, and vice versa. The CSI said Susan was very friendly and talked for about one hour. The whole time they were talking, Susan was holding a blue cosmetics case. The CSI said that it was normal for girls to carry cosmetic cases. During the one-hour conversation, she was open about how back East she enjoyed smoking "weed" (another name for marijuana). She asked the CSI if he cared for some. The CSI told her that he would have to wait until payday. Susan said that she would see him at that time. The CSI later noticed Susan in the male barracks going from door to door, always carrying the blue cosmetics case. I told the CSI I would get back with him.

Meanwhile, the CSI was working nights in a sensitive area when Susan came by to tell him she had something for him. The CSI walked outside his duty section and met with Susan. The CSI told Susan he did not have cash but could give her a check. Susan said that would be okay. The CSI made the check out in Susan's name, and on the bottom left of the check, where it reads "for" and has a line, the CSI wrote the initials *MJ*. The CSI said it stood for marijuana.

I went to the local bank and spoke with the bank manager, who had assisted me in other matters. I asked him to let me know when the CSI's check was received for processing. The CSI did his banking with the base bank, and I figured Susan would cash the CSI's check there.

A couple of days later, the base bank manager called to inform me that the check was cashed by Susan and the CSI could pick it up. I told the CSI to go to the bank, see the bank manager, pick up the check, and meet that evening to release it to me. Later, as planned, I retrieved it under proper receipt of evidence. After coordinating with the staff judge advocate (SJA), he indicated there was sufficient evidence to charge Susan with possession with intent to distribute marijuana.

After updating the base commander, I obtained the authority, in writing, to conduct a search of Susan's barracks room and her vehicle—a 1974 yellow Ford Pinto, license plate number XXX. I finally had Susan come by the OSI. I interviewed her in the presence of Staff Sergeant (E-6) *Wilson (name changed)*, a Security Forces investigator, after advisement of rights. Susan requested a lawyer, and the interview was terminated. I told Susan I had written authority by the base commander to conduct a search of her barracks room and her vehicle.

I first went to the vehicle she had driven to my office, the Ford Pinto. I searched Susan's car with the assistance of Staff Sergeant Wilson. In the front passenger seat was a blue cosmetics case containing a total of eight rolled-up, plastic baggies. They contained a green, leafy substance. The cosmetics case was seized as evidence under proper receipt. The rest of the vehicle was clean. The search of the barracks room did not disclose anything of evidentiary value.

Back at the office, the green, leafy substance was tested and proved positive for tetrahydrocannabinol (THC), the controlled substance in marijuana. My report of investigation (ROI) was forwarded to the hospital squadron commander, a captain, who oversaw personnel problems (unlike the hospital commander, a colonel, who oversaw hospital operations). Susan's squadron commander returned the ROI without action, stating there was insufficient evidence to take any action.

I heard from sources in the hospital that Susan was having an affair with the captain. When the wing commander learned of the captain's indiscretions, and after reading the ROI, he had the captain relieved of his duties and asked the SJA to act. Susan was court-martialed and received a dishonorable discharge.

Weeks later, I received a letter from Susan; she asked me to return her blue cosmetics case. Fortunately, I still had it in my evidence locker. When the SJA said it was okay to return it to her—empty, of course—it was mailed.

CHAPTER SEVEN

THE MASKED ASSAILANT

THE YEAR WAS 1977. It was a winter night, about three o'clock in the morning. I was awakened by a telephone ring from the Security Forces desk officer. He said there was an attempted sexual assault of a female airman, *Jane (name changed),* inside the female barracks. I grabbed my briefcase containing various articles that come in handy at any crime scene and responded to the call.

When I got to the scene, Staff Sergeant *Wilson (name changed),* Security Forces investigator, had already responded. I began to interview the alleged victim, Airman Jane. She said the assailant opened the door to her room; it was unlocked. She said the assailant startled her, so she jumped out of bed. The assailant grabbed her from behind and placed his left arm around her neck, as if he was trying to choke her. As she struggled with the assailant, she could smell the scent of marijuana on his breath. She ultimately bit him on the wrist below his left thumb. When Airman Jane bit the assailant, he released his grip on her, and she screamed as she reached for the light switch. As the assailant ran out the door, she noticed he was wearing a ski mask and a military field jacket. Her room was on the second floor, close to the outside stairs. That was how the assailant escaped.

During the interview of Airman Jane, she felt a small object between her teeth. As she took it out, she said it was a piece of skin from when she bit the assailant. I retrieved as evidence the piece of skin and placed it inside a small bottle from my crime-scene kit. According to Airman Jane, the assailant had not touched anything inside the room except the doorknob, but so had Jane and the other female airmen who responded to her screams.

I instructed the barracks chief to pass the word to lock the rooms at night. Four days later, I received a call from Airman *Clara (name changed)*. She reported someone had been in her room. This happened at about 3:00 a.m., while she was in the latrine (bathroom), which was down the hallway. Airman Clara said the only things missing were some articles of clothing piled on her chest of drawers to take to the laundry room the next day. I got a full description of the clothing: panties, brassieres, sweatshirts, and a sweater.

A week went by, and I received a call from the desk officer after another attempted sexual assault in the female barracks. The time was 3:30 a.m., and when I arrived at the female barracks, Staff Sergeant Wilson was there. This time, there were several girls in front of the alleged victim, Airman *Karen (name changed)*, claiming they had seen the face of the assailant.

I interviewed Airman Karen first. She said she had forgotten to lock her room and was awakened by the masked assailant. She had been reading a book before going to sleep and had left on a small lamp, which allowed her to see that the assailant was wearing a ski mask. Airman Karen said the masked assailant immediately climbed on top of her, placing his hand over her mouth and telling her, if she screamed, he would hurt her.

Fearing for her life, she shook her head no. Airman Karen said her assailant then placed his hands on the headboard and attempted to penetrate her. When the assailant failed to penetrate her, he became enraged and started cussing and using the F-word as he got off her. As the assailant moved towards the door, he swung it open extremely hard. Airman Karen said that was when she started screaming.

When she stepped out into the hallway, she could see the other girls coming out of their rooms. They ganged up on the assailant and started throwing punches. During the confrontation, the girls managed to remove the ski mask. Airman Karen said she ran to help the other girls, but the assailant managed to escape from the girls; he ran down the second-floor hallway. She said one of the girls caught up with the assailant at the top of the stairwell and pushed him down the stairs. The assailant got up and ran into the night.

I asked who had seen the assailant's face, and five girls raised their hands. I requested they come see me later that morning. When they arrived, I asked each one of the girls to assist me in developing a composite of the assailant. I accomplished this task, in the presence of our secretary, using the OSI Identi-Kit. When I completed the five composites based on the descriptions provided by the five witnesses, I was amazed at the close resemblance of each composite.

The following day, I was in luck—all the first sergeants of all the organizations were holding a meeting. I attended and asked them if after their meeting they could stop by my office and look at some composites I had on a table. One first sergeant stopped by and after looking at the composite said, "He is a dead ringer for one of my unit members. That would be Airman *Jackson (name changed).*"

I did not waste any time. I had Airman Jackson come to my office. As I advised Airman Jackson of his rights, he requested a lawyer. Before I released Airman Jackson, I fingerprinted him, concentrating on his left wrist area below the thumb, where a piece of skin was missing. I also obtained the authority, in writing, from the base commander to conduct a search of Airman Jackson's barracks room. In the barracks room, I noticed a military-green field jacket. I was more interested in finding a ski mask.

I opened the top drawer of a chest of drawers. To my surprise, I discovered articles of ladies' clothing: panties, brassieres, sweatshirts, and a sweater. I placed the clothing in a plastic bag and seized them as evidence. I opened the second drawer and found a ski mask. I seized the jacket and ski mask as evidence under proper receipt.

The following day, Airman Clara came by the office and provided a written statement wherein she positively identified the articles of clothing as hers. I also forwarded the latent print I lifted and the piece of skin from Airman Jackson's scrape on his wrist to the Army's criminal laboratory for comparison.

In the meantime, Airman Jackson's lawyer from the Area Defense Council (ADC), in coordination with the staff judge advocate (SJA), requested a lineup with Airman Jackson. I assisted with the lineup by locating four people with similar builds, color of hair, and weight, and they would all be wearing a military-green field jacket. The lineup was set with Airman Jackson in the fourth position. The lineup was in the courtroom, and each of the five female witnesses was escorted into the courtroom. When a witness would make an identification, she would say, "I'm ready," and be escorted to another room where the SJA prosecuting attorney and Airman Jackson's defense lawyer were waiting. All five female witnesses picked number four, Airman Jackson, as being the assailant.

Airman Jackson was arrested by Security Forces and placed in pretrial confinement, pending court-martial. In the meantime, I talked to the laboratory technician who said, although the outline of the piece of skin and the latent print were similar, he could not say definitively they came from the suspect because the piece of skin had withered in the bottle. Formaldehyde would have saved the sample. I said I would remember that the next time.

Airman Jackson was court-martialed, dishonorably discharged, and had to serve four years at Fort Leavenworth, Kansas.

Photo: Courtesy of U.S. Air Force

In the photo above, I (right) use an Identi-Kit to construct a composite based upon a description of a subject provided by a witness. This same process was used to construct a composite of the assailant who accosted some female airmen in their barracks.

CHAPTER EIGHT

DIRT-BIKE TRAILS

THE YEAR WAS 1978. On a nice spring morning, I received a call from the base commander who asked me to meet him on the number six green at the base golf course. I never asked any questions when the boss called. I arrived at the golf course clubhouse, jumped on a golf cart, and headed to the number six green. When I arrived, there were a couple of Security Forces (SF) patrol vehicles and the base commander close to the green.

Preliminary examination disclosed that, on the green, there were tire marks made apparently by a motorcycle(s). There were a lot of circular tire tracks that had literally dug up the green. There were also tire tracks in the front sand trap. With the early morning dew on the grass, it appeared the suspect(s) raced the motorcycles from the fairway into the front sand trap and jumped the motorcycle(s) onto the green, causing additional damage. I asked the patrol officers to look after the crime scene while I went to my office for some plaster of Paris for the tire tracks. I then looked at the base commander, who was an avid golfer, and told him I did not have to brief him on this incident. I had played golf with the base commander on a few occasions (see chapter 19, "My Amusing Tidbits").I did ask if he was requesting an investigation.

He replied, "You find the people responsible."

When I later returned to the crime scene, I made an impression of the tire tracks. I collected samples of the particles of grass left behind on the golf green. Golf course maintenance personnel were standing by to immediately create a temporary green (those who play golf know it is not the same as real grass). My next move was to visit the barracks on base that had motorcycle parking areas.

At the second barrack's motorcycle area I visited, I found two dirt bikes parked next to each other. I noticed particles of grass debris on their tires. Both engines were still warm. I got written authority from the base commander to seize both dirt bikes. I requested Security Forces to bring their flat trailer to transport the dirt bikes to the vehicle impound lot for further examination.

Through researching the base registration records, I discovered the dirt bikes belonged to Airman *Jones* and Airman *Walker (names changed)*. I again did not waste any time. I had both Jones and Walker come by my office. I first interviewed Airman Jones, and after advisement of rights, he did not request a lawyer. When I asked Jones to tell me about riding his dirt bike early that morning, he replied with the whole story.

He and Airman Walker recently purchased the dirt bikes and decided to ride south of the base housing in the brushy area where there were trails for dirt bikes. They were riding along until they came across an open grassy area. They noticed an open sandpit where they could race their bikes and make them jump over the lip of the sandpit and onto the smoother grassy area. That was where they could easily raise the front of the bikes, spin on the rear tires, and make circles. They both completed this maneuver and headed back to the barracks. I asked Airman Jones if he knew he was on a golf course. Jones replied he did not and had never been on a golf course before.

I then interviewed Airman Walker, who after advisement of rights did not want a lawyer. Airman Walker essentially gave me the same information as Airman Jones. Both Jones and Walker provided written statements.

I then provided my report of investigation (ROI) to Jones's and Walker's unit commander for action. The unit commander, in coordination with the staff judge advocate (SJA), decided to administer an Article 15 under the Uniform Code of Military Justice (UCMJ), which is a disciplinary action less than a general court-martial. Both Jones's and Walker's driving privileges were revoked for six months, and they were ordered to pay damages to the golf course. According to the base civil engineers, the bill for repairs was $8,000.

The base newspaper carried the news of how two airmen, who were thrill-riding on dirt bikes, caused damages to a golf course green and were ordered to pay $4,000 each.

CHAPTER NINE

JET ENGINE FOR SALE

THE YEAR WAS 1979. I had a close friend who was a criminal investigator for the state of Texas. On a summer day, he reported that a colleague in the narcotics investigations division had a confidential source of information (CSI) who had certain information about a female airman, Airman *Brown (name changed)*. I immediately wanted to know about the airman.

I met with the state narcotics agent who related his CSI had been at a party where marijuana was being smoked. The CSI was getting to know Airman Brown well. She told the CSI that she worked on jet engines at the airbase. She said security was so relaxed at the hanger where the jet engines were kept that she could easily take a pickup truck and drive away with one. The quick-thinking CSI created a story about a businessman he knew who had connections in a foreign country and would probably be interested in the engine. The CSI added the businessman would be willing to pay a good price.

I told the state narcotics agent I needed to check out Airman Brown. I just happened to have another CSI who worked on the jet engines. My CSI said that Airman Brown did, in fact, work on jet engines in the hanger where they awaited maintenance. The jet engines were on powerful hydraulic jacks with heavy-duty wheels, so they could be easily moved around. The jack could raise and lower the engines, certainly making it easy to lift one and place it on a pickup truck. My CSI said that the hanger was well lit at night, and Security Forces' patrols checked the hanger about once every hour.

With the information I had, I briefed the base commander, who requested an investigation. I met with my close friend (the state criminal investigator) and his colleague; I asked the narcotics agent if his CSI could continue to meet with Airman Brown and work a deal with her. A few days later, the narcotics agent's CSI called and said he had a deal with Airman Brown. On the condition that she could see the money first, Airman Brown was willing to steal the jet engine for $25,000.

I had to coordinate with my District Headquarters and go over all the details. The local wing commander, who was responsible for all flying operations on base, wanted to resolve this situation as soon as possible. It took a couple of days for District Headquarters to borrow the $25,000 in cash from the accounting and finance office. Then, we needed an OSI undercover agent (UCA) to portray the businessman who had the foreign connections. A very experienced OSI agent with grey hair was perfect for the part. The next step was to get the narcotics agent's CSI to set up the meeting with Airman Brown and show her the money.

The day came for the meeting. I asked a couple of federal agents if they could cover the OSI UCA, who was also armed, during the transaction. After the OSI UCA showed Airman Brown the money, she indicated that everything was in order; she was ready to complete the job and deliver the stolen jet engine. Airman Brown told the narcotics agent's CSI that she needed a pickup truck. The engine delivery would be at the specified marina of a nearby lake. The state CSI borrowed a pickup truck from the state agency that had seized one in a narcotics raid. He drove the pickup truck to the base to pick up Airman Brown.

It was about midnight when Airman Brown and the state CSI drove to the flight-line hanger. Some federal agents who wanted to help were standing by at the marina, waiting for the delivery of the jet engine. However, some bad luck fell upon us. The pickup truck broke down right before reaching the hanger. The engine theft did not happen. The state CSI walked Airman Brown back to her barracks and then departed. I contacted the folks at the lake marina to let them know what happened.

When I briefed the wing commander, he said he had heard enough about Airman Brown's intentions. He asked the staff judge advocate (SJA) to facilitate a discharge for Airman Brown. Eventually, the SJA arranged a "general court-martial with judge alone."

Brown received a dishonorable discharge. The $25,000 was safely returned to the accounting and finance office. The wing commander, after discussions with his flight-line commanders and superintendents, tightened up security around the jet engines awaiting maintenance in the hanger.

CHAPTER TEN

AIR FORCE BEEF UNINSPECTED

IN THIS CHAPTER, I illustrate that one of the key elements in conducting criminal investigations is the development of sources of information. I always aggressively pursued source information for two reasons: first, to determine the reliability of the source, and, second, to explore the information itself.

The year was 1978. I received a call from a confidential source of information (CSI) who worked in the meat department of a military commissary. The CSI related that he was told by a friend in the meat processing plant in Laredo, Texas, that cattle from Mexico was being delivered to the plant, which was legal. However, the friend said the meat was being processed without proper inspection by the United States Department of Agriculture (USDA). It was being stamped "USDA," but not inspected.

I traveled to Laredo, Texas. Fortunately, on the way, I stopped to see a friend in the local FBI office. During our conversation, I mentioned what I was following up on. My friend recommended I first contact my District Headquarters commander. I did and was told to see him before I followed up on the information. I was genuinely concerned about what I had stumbled across.

When I got to District Headquarters, I went to see the commander, whom I considered a mentor. I always had good conversations with him. He finally told me I had to stop following up on the information in Laredo. He hesitated in telling me why. When he noticed my disappointment, he said, "The Bureau has it." Of course, he meant the FBI. I thanked him and happily went on my way.

I later contacted my CSI and told him the information was good, but someone else was looking at it, and I would have to wait awhile. About thirty days later, *Newsweek* magazine carried the story of how the FBI had infiltrated the Small Business Administration (SBA) and had arrested ten SBA ranking officials in the Southeast region of the United States. It seemed that the SBA officials were receiving payoffs from contractors (like meat-processing plants). When the story broke in *Newsweek* magazine, I presented a copy to my CSI at the military commissary.

CHAPTER ELEVEN

GARDENING MARIJUANA

THE YEAR WAS 1978. During the interviews of several airmen in a narcotics investigation, I came upon an Airman *Baker (name changed)* who had about one month left in his Air Force four-year commitment. Since he was leaving the area, I asked Baker for the name of the individual dealing marijuana on base. He said he would tell me only if I could arrange for him to depart immediately. I told him I would speak to Military Personnel.

I spoke to the base commander and the chief of Military Personnel. Both agreed to release Baker early if I got what I wanted. I got ahold of Baker to let him know we had a deal for early release. I asked him to give me the dealer's name, and as soon as I could confirm it, Baker would be released.

Baker gave me the name of Sergeant *Jack (name changed),* who was growing marijuana plants in the back of his house off base. Baker said Sergeant Jack had around thirty or forty plants, all about five feet tall, and was hoping he could harvest a few pounds of "weed" (marijuana). Baker said Sergeant Jack could be holding additional marijuana inside the house and that he owned a .357 Magnum pistol. Baker provided Sergeant Jack's home address and physical description of the house. I told Baker that I would check out the information and get back with him.

I went to see the local district attorney's chief investigator, who assisted me in obtaining a search warrant for Sergeant Jack's off-base residence and the surrounding premises. The local law enforcement officers and I executed the warrant when Sergeant Jack was home. We immediately went to the rear of the house where thirty-five, five-foot-tall marijuana plants were removed from the premises. Officers estimated between five or six pounds of marijuana could have been harvested.

Then, the search of the inside of the residence began while a police officer watched over Sergeant Jack in the living room. Sergeant Jack asked permission to go to the bathroom. Since I was standing close to his bedroom and bathroom, I said I would watch him. Sergeant Jack walked in front of me into the bedroom. As he was making the turn at the foot of the bed, I quickly noticed a firearm on top of the nightstand. I immediately reached for my firearm and yelled out, "STOP! Don't even think about it." I then escorted Sergeant Jack back to the living room. I was wearing surgical gloves when I handled the weapon on the nightstand. It was a .357 Magnum; I unloaded the weapon to later determine to whom it was registered.

The search of the house only disclosed a kitchen drawer containing debris of a green, leafy substance, plastic baggies, and a small scale. No other illegal drugs were found. The debris in the drawer tested positive for THC, the illegal substance in marijuana. Sergeant Jack was arrested by the local police, but when he appeared before a judge for arraignment, the judge released him to the custody of the military.

Sergeant Jack was placed in pretrial confinement by Security Forces at the airbase. A county-records check of Sergeant Jack's residence revealed that the house deed was in his wife's name. That changed the ownership of the evidence and the person responsible for the contents in and around the house premises.

While Sergeant Jack was in pretrial confinement, he asked to see me. I had not officially interviewed him, so I was cautious. I wanted to just listen first. He told me that he had heard that I treated people with respect, and he wanted to ask a favor. It was about his wife and three-year-old daughter. He said they had nothing to do with what was found in their house.

I stopped Sergeant Jack and told him I had to advise him of rights, but he did not want a lawyer. Sergeant Jack was willing to provide a written statement of admission. I told him I would talk to the staff judge advocate (SJA).

The SJA said the military had no interest in Sergeant Jack's family. He was happy with the military's position. In the court-martial, Sergeant Jack was represented by legal counsel but pled guilty. He received a bad-conduct discharge and had to serve twelve months at Fort Leavenworth, Kansas. Incidentally, the weapon in Sergeant Jack's house was legally his and was returned to his wife.

CHAPTER TWELVE

DEATH OF THREE SOLDIERS

It was 7 December 1977. I received a telephone call from *Comandante Perez (name changed)*. In Mexico and Latin America, the word *comandante*, or commander, is used to address a person in high office of either the military or law enforcement (in this case, chief of police). Comandante Perez indicated that his officers had discovered the bodies of three United States Army soldiers in a local motel room, apparently from asphyxiation by gas-fume poisoning.

The next day, December 8, I visited Comandante Perez in the Mexican border town of Ciudad Acuña. After discussing the deaths of the three U.S. Army soldiers, Comandante Perez offered to show me the motel room where the three soldiers were found. I made it noticeably clear that I was not there to conduct any investigation. The comandante acknowledged my purpose there and arranged for a couple of officers to take me to see the motel room.

I made the following personal observations: There were two twin beds and a sofa. The room smelled of liquor and marijuana. There were three tequila bottles (almost empty) and some marijuana "roaches" (marijuana cigarette butts) in the ashtrays. Apparently, the soldiers smoked marijuana and drank tequila to the point of falling asleep. At that time, the U.S. Army paid soldiers at the end of each month, so since it was only December 7, the three soldiers had to have been carrying some money and had probably driven to the border town. I asked the two officers if they had found any money on the soldiers. The officers said there was no money, just the soldiers' military cards and some photos. The officers also said they had found no car belonging to the soldiers. Since it was wintertime, there was a space heater in the room. The flames in the space heater could have gone out, filling the room with gas fumes; however, the money and the soldiers' car were gone.

I finally asked the officers what they suspected. The officers thought it was possible that the manager of the motel waited until the three soldiers got drunk and fell asleep before he turned off the gas from outside the room to extinguish the flame and turned it back on to fill the room with gas fumes. The officers said the motel manager disappeared when the three soldiers were discovered. The officers suspected the motel manager took the soldiers' money and car, then disappeared into the interior of Mexico.

I returned to Comandante Perez's office, where I asked whom I had to see about returning the three deceased soldiers' bodies to the United States. I was told to go see a lawyer named Flores in Public Ministry. I did not speak to the media or the reporter who took our picture in Flores's office. I had read newspaper articles from Mexico and Central America on occasion, and I knew that, typically, stories are sensationalized by the media in those countries. The locals believed the three soldiers were from Houston, Texas. Meanwhile, the local coroner had already processed the soldiers' remains. The bodies were transported to a funeral home in Del Rio, Texas.

I knew there was no U.S. Army installation in Houston, Texas. However, the nearest Army post to Houston was Fort Hood in Killeen, Texas. I contacted the Army's Criminal Investigation Detachment (CID), who were able to track down the soldiers' unit after I provided their names. Fort Hood's Mortuary Affairs officer planned to have the soldiers' remains transported from Del Rio to Fort Hood.

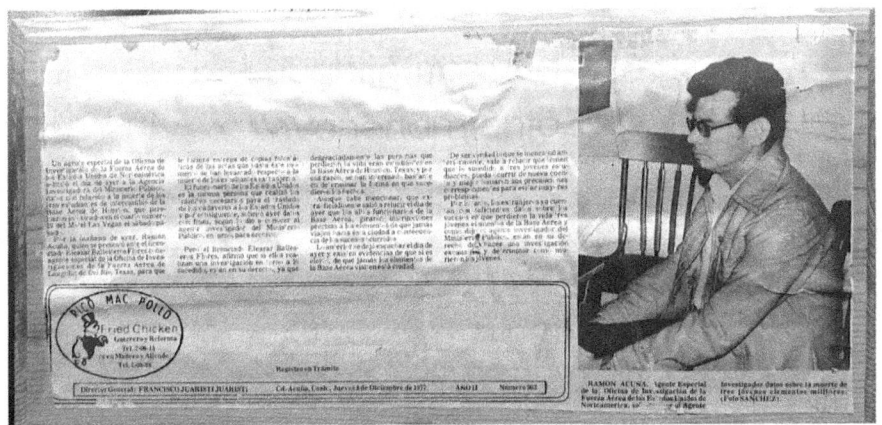

Article and Photo: Author negotiating the return of the deceased soldiers' bodies to the US.

CHAPTER THIRTEEN

IMPERSONATION AND MURDER

THE YEAR WAS 1973. A call came into the OSI from a first sergeant who suspected an airman in his unit of impersonating an Air Force major to impress the airman's girlfriend. The first sergeant further identified the individual as Airman *Adams (name changed)*. Being a rookie agent, my boss assigned me this case.

I was ready to interview the suspect, who denied the allegation. I was also prepared to interview the suspect's girlfriend and close friends at work, whom he had bragged to about the impersonation. The day Airman Adams came to my office, he showed up with his girlfriend in tow. I thought to myself, *How convenient is this!*

After I advised Airman Adams of his rights, he did not want a lawyer. He said he had been dating his girlfriend, *Peggy (name changed)*, for about six months. She had not seen him in uniform, so he told her that he was an officer. Airman Adams said one day he went to Peggy's apartment wearing the major's uniform; they went to a store and returned to her apartment. Airman Adams said that was the extent of the impersonation, and he had only worn the major's uniform one time. Airman Adams agreed to provide a written statement. I asked him how old he was, and he replied he was twenty-one years old. I said that was too young to be a major.

I told him I was going to talk with Peggy. I interviewed Airman Adams's girlfriend, Peggy, in the presence of the OSI secretary. Peggy provided essentially the same details as provided by Airman Adams and was willing to provide a written statement. She hoped Airman Adams did not get in trouble. I told Peggy that there were rules in the military against impersonation; however, it would be up to Airman Adams's commander whether or not to take further action.

Just one week later, while Airman Adams's commander was still pondering what action to take about the impersonation, I received a call from the Security Police desk officer. He stated that the base motor pool had reported a Lieutenant Adams checked out a station wagon early one morning and was to return it the same day. When motor pool personnel attempted to contact Lieutenant Adams's duty section, they were told they had no Lieutenant Adams, but they did have an Airman Adams.

My curiosity was building up as I went to the base motor pool. I talked to the personnel who checked out the station wagon to a Lieutenant Adams; their description of the lieutenant matched my Airman Adams. I immediately contacted Airman Adams's girlfriend; she had not heard from or seen him in the last few days. She told me she would call me if she heard from or saw him.

A couple of days later, the local TV newscast reported the holdup of a local pawnshop; the proprietor was shot and killed. An eyewitness saw four young men run out of the pawnshop and get into a blue station wagon with the words "U.S. Air Force" on the side. I instantly contacted the San Antonio Police Department (SAPD) and spoke with the detectives working the pawnshop-killing case. I related the details of my case on Airman Adams. I also gave them his girlfriend's name and address. I was certain the SAPD was going to need to talk to Peggy.

A couple of days went by, and then I got a call from Detective *Art (name changed)*, SAPD, who advised that the four young men in the San Antonio pawnshop killing had held up another pawnshop in Houston, Texas. Detective Art said that a witness indicated four young men used a blue station wagon to make their getaway. I subsequently got a call from Peggy, who informed me Airman Adams had called her from a local hotel where he was visiting friends. I contacted Detective Art.

The next day, I got a call from Detective Art; he told me they arrested the four suspects at a local hotel. The blue Air Force station wagon was still in their possession. He invited me to come by his office and look at photos of the four suspects. I identified Airman Adams as one of the four.

I noticed another one of the suspects had a couple of black eyes. I asked, "What happened to this guy; did he run into a door?"

Detective Art said, "Not exactly." It turns out, during the arrest, the suspect had run into a Texas Ranger who had volunteered to help them on the case; he stood six foot five inches tall and weighed 200 pounds.

All four men were prosecuted by the Bexar County District Attorney in San Antonio, Texas.

CHAPTER FOURTEEN

THEFT OF MILITARY PROPERTY

IN THIS CHAPTER, I will first set the scene for the investigation that follows.

The year was 1979. The military services developed two new organizational concepts for base operations. It involved the consolidation of civil engineering and contracting functions. The concepts were to be field-tested for five years in San Antonio, Texas, on five existing military installations: Randolph Air Force Base (RAFB), Brooks Air Force Base (BAFB), Kelly Air Force Base (KAFB), Lackland Air Force Base (LAFB), and Fort Sam Houston (FSH).

Under the consolidation concepts, the engineering function became the San Antonio Real Property and Maintenance Agency (SARPMA). The contracting function became the San Antonio Contracting Center (SACC). SARPMA was commanded by an Air Force officer, a colonel (an O-6), and SACC was directed by a civilian government employee (a GS-15). Both SARPMA and SACC were located at FSH because of space availability.

In 1981, I was assigned to the OSI Team at RAFB and was responsible for covering SARPMA and SACC. I soon received a call from a confidential source of information (CSI) who related that SARPMA had recently been inspected by Headquarters Air Force personnel, and one of their findings was that SARPMA was overstocked in materials and supplies. The dollar estimate of the overage was more than one million dollars.

To reduce the inventory, the CSI said that SARPMA had already begun to turn in materials and supplies to the Defense Property Disposal Office (DRMO) located in a huge warehouse on FSH. Fort Sam Houston's DRMO branch would then transfer most of the materials and supplies to the main DRMO depot at Kelly AFB. The depot accepted turned-in government property from several Southern states. The CSI stated that the turn-in document for government-like items came in several copies to show the receipt, transfer, or release of the same property.

FSH DRMO was being managed by Mr. *Charles* (name changed). The CSI said he had seen property received with turn-in documents signed by Mr. Charles, which meant a reduction in the SARPMA inventory. The signed turn-in documents had been steadily coming in for a few weeks. However, when the CSI then called the DRMO depot at Kelly AFB to verify they had received the inventory from some of the first turn-in documents signed by Mr. Charles, he found that none had been processed at Kelly AFB. According to the CSI, the dollar value of the "missing" SARPMA property was approaching $500,000.

After telling the CSI to let me find out what was going on, I immediately contacted a CSI at the DRMO depot, KAFB. He had assisted the OSI in the past and was a warehouse worker at both locations, FSH and KAFB. The DRMO CSI said he would be working the next day at FSH for a couple of days. I asked the DRMO CSI to find out what Mr. Charles was doing with the turn-in documents of the SARPMA property.

The first day of the CSI's assignment, he contacted me and said he noticed several turn-in documents in two trash cans that were full of forms. The trash cans were next to a desk in Mr. Charles's office, but the CSI could not get the turn-in documents because it would look suspicious. He did notice Mr. Charles empty those trash cans into the dumpster next to the warehouse building.

That evening, I went "dumpster diving." When I opened the lid of the dumpster, I readily spotted a collection of turn-in documents. I proceeded to gather all I could see. Later, after examination of the turn-in documents, I discovered they all reflected Mr. Charles's signature. From each set, the original copy was missing—it would have gone to SARPMA. The remaining copies were intact.

The next day, I got a call from the DRMO CSI and was told that Mr. Charles had ordered that most of the new building materials be loaded onto a one-and-a-half-ton truck. Mr. Charles appeared very friendly with the truck driver and passenger. Later, Mr. Charles would say that they were government employees working on a project and that DRMO depot had given him the okay to release the property.

The CSI provided the license plate number of the truck and informed me that they would return in two days. That gave me enough time to get the OSI District Technical Services Division to set up technical surveillance in the warehouse building next door to the loading dock of DRMO.

Two days later, the same truck with the identified license plates backed up to the loading dock. This time, what followed was caught on film. Again, the truck was completely loaded with construction material. Same driver and passenger. This time, I was also ready to follow the truck. I employed every trick possible to remain unseen while following the truck. I traveled ten miles before the truck stopped. It was a lumberyard near the outskirts of the city. Workers began to unload the truck. I had seen enough.

The next day, I met with agents of the local FBI office. I briefed the FBI agents on the details of this investigation. They also viewed the surveillance film of the truck being loaded while Mr. Charles, the truck driver, and his passenger stood and waited. The FBI agents said, the next time the truck showed up, they were ready to arrest Mr. Charles, the truck driver, and his passenger; they were also going to check out the proprietor of the lumberyard.

The following day, the CSI from the DRMO was back at the FSH branch and promptly called to say, "They are here."

The FBI agents responded instantly. They said that they would arrest Mr. Charles as soon as the truck departed. When they truck was a block away, I was cleared to stop it and take the driver and his passenger down. That made my day.

When the truck was loaded and started to depart, I took off, went about one and one-half blocks, stopped, and waited for the truck. I positioned my red Kojak emergency light on the roof of my car and used the car to block the road. I stood behind my open car door and directed the driver and passenger to exit the truck with their hands up, but they hesitated. That is when I drew my weapon, a .45 Combat Master Piercer. They both jumped out of the truck with their hands up. The FBI agents arrived moments later and took over. The truck was seized as evidence.

I later learned that Mr. Charles confessed to receiving money from the lumberyard proprietor. The reason Mr. Charles, a sixty-year-old man, was making extra money was to support his twenty-nine-year-old mistress. The FBI agent said the truck driver, his passenger, and the lumberyard proprietor also confessed.

CHAPTER FIFTEEN

PAYMENTS TO A GHOST

IN THIS CHAPTER, I illustrate that there is no limit to an OSI agent's imagination—for example, when evaluating the information received from complaints via telephone calls or walk-ins.

The year was 1983. I answered the telephone call from an anonymous female caller. She stated that she had gotten to know, in her neighborhood, a woman who had been married to a retired military veteran. The anonymous caller continued to say that, although the military retiree had been dead for several months, the widow continued to receive the retiree's pay. I then asked the anonymous caller for the name of the widow and her address. The widow's name was *Mrs. Green (name changed)*, and her address was provided. I thanked the anonymous caller for the information and told her that I would follow up on the matter.

I knew that Air Force retiree payments were centralized at the Air Force Accounting and Finance Office (AFAFO) in Denver, Colorado. Also, all Air Force retiree deaths had to be reported to the Veterans Administration (VA) in Washington, DC, so that their military benefits could be stopped or adjusted. I did not open an investigation into this matter. However, I did initiate a fraud information report (FIR) and recommend sharing the information with the nearest OSI detachment to the AFAFO in Denver, Colorado, and the VA in Washington, DC, in order for them to follow up on the information. Coincidentally, to provide forensic consultations to field agents, the OSI had recently hired some officers with computer systems backgrounds.

Agent *Al (name changed)* assisted agents in Denver and DC to obtain the correct computer products to merge databases and determine if the Air Force was still paying retired veterans who were dead. He eventually learned that the Air Force Accounting and Finance Office did not really conduct good oversight in the matter of military-retiree payments or coordinate well with the VA to determine if retirees had expired. The AFAFO assured the OSI that they would tighten their procedures and coordinate better with the VA.

The anonymous tipster later called back and said the neighbor was no longer getting payments for her deceased husband. In any case, bingo for the OSI agents who are truly Guardians of Air Force Resources.

CHAPTER SIXTEEN

KOREAN MARRIAGE SCAMS

THE YEAR WAS 1981. A CSI (confidential source of information), while stationed in Korea, contacted me with some interesting information. The CSI said he was approached by a Korean national at the base noncommissioned officers' (NCO) club the previous night with a proposal. The Korean national asked the CSI if he wanted to make $3,000 "very easy." The Korean gentleman explained that all the CSI had to do, with the assistance of certain prearranged individuals at the American embassy, was marry a Korean female and get the female a visa. Again, with the help of certain embassy individuals. Once the Korean female traveled to the United States as a dependent wife of an American military member, the CSI would never see or hear from the Korean female again. I told the CSI to give me an opportunity to coordinate this information with command and for him to call me later.

In discussion with the base commander (BC) and the staff judge advocate (SJA), they both expressed interest in how many military members had fallen for the marriage scheme. The BC requested an investigation.

I then received another call from the CSI, and I asked him if he was willing to go through with the marriage scheme and be monitored by the OSI? The CSI agreed.

The CSI was again contacted by the Korean gentleman, whom I now called the "arranger." The CSI told the arranger he would go through with the proposal. The whole scenario was set, and I asked for assistance from other OSI agents to help monitor the activities of the CSI as the scheme unfolded.

The day came when the CSI was introduced to the Korean female, who spoke some English. The CSI was careful to remember the people he had to see, the ones who would perform the marriage and take care of all the paperwork. The marriage was conducted, and then the CSI proceeded to the section at the US Consulate where he could obtain a visa for the new wife of an Air Force member.

The next day when all the paperwork was completed, the arranger met with the CSI and paid him $3,000. So, as the CSI and his bride were set to arrive in Osan Air Base, where the new bride would board an airplane destined for the United States, they were met by OSI agents.

Subsequent interviews of the CSI's new bride revealed she had paid the arranger a total of $10,000 for her travel to the United States. We were able to identify those people the CSI had contacted in the consulate, and all Korean nationals involved were relieved of their duties. We did learn that, of the most recent marriages performed, there were thirteen Air Force members involved. The arranger was picked up by the Korean National Police, who indicated they had gotten a confession from the arranger; he admitted paying money to those contacts in the United Stated Consulate. The CSI's new wife agreed to a marriage annulment, and the CSI returned the $3,000 to the female Korean.

Back at the base, we had thirteen Air Force members to interview. The grades of those airmen went from technical sergeant (E-6) to sergeant (E-4). I first interviewed the technical sergeant who, after advisement of rights, provided a written statement wherein he admitted to taking part in the marriage scheme. However, included in the advisement of rights was a charge of bigamy since he was already married back in the United States. The technical sergeant argued the marriage was not real; he was never going to see or hear from the Korean female he married. Regardless, according to the SJA, the marriage at the consulate was real, and the offense was bigamy. Although bigamy is not a court-martial offense, it was up to the technical sergeant's unit commander to take some sort of action under Article 15.

The remaining twelve airmen, who were all married back in the US, provided written statements essentially with the same information that the technical sergeant had given. They were all told the same thing about the marriages being real. They were told the charges against them were not court-martial offenses, but their unit commanders could elect to take administrative actions against them.

Once the thirteen airmen returned to the US, they may or may not have told their wives about their Korean marriage schemes and the money they received. Then again, that is a big secret to hide.

Note: In 1981, the details of this investigation were shared with the National Security Agency (NSA) as a *modus operandi* for entering the United States.

CHAPTER SEVENTEEN

CRITICAL DRUG INVESTIGATIONS

IN THIS CHAPTER, I address two separate investigations: one at Laughlin AFB in Del Rio, Texas, and the other at Holliman AFB in Alamogordo, New Mexico. The common link between these two investigations was the military personnel involved. They were assigned to communication squadrons at both sites. A large number of personnel in those units performed their duties in radar-approach facilities. Their functions are similar to those performed by civilian air traffic controllers at civilian airports. The military members track military aircraft, when they are approaching the runway, and try to ensure safe landings and takeoffs, especially in inclement weather conditions. It is imperative the personnel have clear minds when performing their duties. No one wants these specialists to sit in front of the *Star Wars*-like screens with fuzzy minds.

At Holliman AFB, I assisted a team of OSI agents with an expanded narcotics investigation. Not only were the separate investigations targeting two like units, but ultimately, at the conclusion of both cases, the court-martials that followed were taking place the same week. These became a challenge for me, but not for the Air Force. At the end of this chapter, there is a photo of me that shows how the Air Force accomplished this mission.

Now, I will start with the investigation at Laughlin AFB that I initiated. My case began when a confidential source of information (CSI) wanted to talk to me. I met the CSI; he gave me the accounts of Airman *Echo* and Airman *Franks (names changed)*, both assigned to the communications squadron. According to the CSI, both Echo and Franks were supplying marijuana to unit personnel. Additionally, they had run dry and were getting ready to go pick up some more marijuana from the Rio Grande. The CSI indicated Airmen Echo and Franks had talked about traveling to Big Bend National Park on their motorcycles and visiting the banks of the Rio Grande. The CSI believed this is where they could score (buy) the marijuana. I asked the CSI if he could find out when Echo and Franks would be making the trip. The base commander was briefed on the details of this matter and requested an investigation.

In the meantime, I alerted the two customs patrol officers (CPOs) who patrolled the Big Bend National Park area. A couple of days later, on a weekend, my CSI called and said Echo and Franks had departed on their motorcycles and were heading to Big Bend National Park. (Note: It is about five hours, one way, from Laughlin AFB to Big Bend.) I immediately notified the Big Bend CPOs, and I was certain they would have good means of surveillance.

The CPOs later told me their good friend, a Texas state trooper, noticed two motorcyclists enter the only road to Big Bend as he took off in the opposite direction. After he radioed the CPOs, the CPOs began their surveillance of the two motorcyclists. They surveilled them going to the banks of the Rio Grande, where the CPOs witnessed the two motorcyclists meet up with a small boat. A man in the boat handed the two motorcyclists a large sack. One of the motorcyclists handed the man what appeared to be some money.

The CPOs waited until the motorcyclists loaded the sack on one of the motorcycles, and as they were exiting the park, they were stopped and arrested for possession of marijuana. Both airmen were arraigned and appeared in front of a judge in the small town of Mathis, Texas, at the entrance of the only road to Big Bend National Park. Both airmen were released on their own recognizance. However, back at the base, bigger things were waiting for them.

My CSI soon contacted me and said both Echo and Franks were back in Laughlin AFB; they were both unhappy campers with no marijuana. I quickly wrote my report and forwarded it to Echo's and Franks's unit commander for action to be taken. While the unit commander and the staff judge advocate (SJA) worked on scheduling a court-martial, I was tasked to travel to Holliman AFB in New Mexico to assist in their extended drug investigation for one week.

While the case at Laughlin AFB involving Echo and Franks was pending trial, I got involved with the case at Holliman AFB in New Mexico. When I arrived at Holliman AFB, the team of OSI agents were ready to conduct a lot of interviews and searches of barracks and residences. The local OSI detachment had a CSI who had made a couple of control buys of cocaine from a technical sergeant (E-6); the CSI had also identified some marijuana dealers and users of marijuana and cocaine. As the interviews proceeded, some wanted lawyers, but others did not. Those who did not were asked for consent to search their barracks room. Of the barracks rooms searched, I did not find any marijuana.

On my last day at Holliman AFB, I interviewed a sergeant (E-4) who was suspected of dealing marijuana. He did not want a lawyer as he denied the allegation and consented to a search of his off-base residence; he was married. When I went to the suspect's residence, I met up with two detectives from the Alamogordo Police Department (APD); they had volunteered to assist the Holliman OSI team in their drug investigation. The three of us were greeted by the suspect. Initial impression of the modest residence was its immaculate cleanliness. The two detectives searched every corner of the residence, but found no marijuana or even the scent of marijuana.

As the two detectives and I were ready to depart, we went into the living room, where the suspect and his wife were sitting on the couch. As I was complimenting the suspect's wife on the neat appearance of the house, I walked up to an end table with a lamp sitting on it. The end table had four panels around it. I lifted the top part, and it moved, almost knocking the lamp down. I removed the lamp and lifted the top of the end table; the scent of marijuana was strong. Inside the end table was a large amount of marijuana, a small box of sandwich baggies, and a small scale.

The marijuana weighed one pound and fourteen ounces. The suspect was arrested by the detectives and was transported to

Holliman. I provided the suspect's wife his post commander's name and number so that she could keep informed about any action taken. The suspect was released to the security forces at Holliman.

As I prepared to depart Holliman AFB, I was told I would be contacted later to testify in the suspect's court-martial. That meant I was going to be testifying at both Laughlin AFB, Texas (my home base), and Holliman AFB, New Mexico.

Then, it finally happened. The court-martials of Airmen Echo and Franks at Laughlin AFB were scheduled for the Thursday and Friday of a set week, and I was contacted by Holliman that the court-martial of the sergeant with the marijuana was scheduled for Monday and Tuesday of the same week. Flying commercially from Laughlin was a big challenge for me because there is no direct flight. I would have to drive to San Antonio and then catch a plane to El Paso, Texas, where I needed to rent a vehicle and drive over 200 miles to Alamogordo, New Mexico. That would mean too much time consumed in travel time in order to attend the two court-martials.

However, it was a simple mission for the Air Force. I flew in the back seat of a T-38 trainer. The pilot and I departed Laughlin on a Sunday. I testified on Monday/Tuesday and flew back from Holliman on Wednesday, in time to testify in Laughlin on Thursday/Friday. Incidentally, the pilot earned his cross-country hours for the mission.

MEMORABLE MOMENTS/AWARDS

Photo: Courtesy of U.S. Air Force

The only way to fly. I flew from Laughlin AFB in Del Rio, Texas, to Holliman AFB in Alamogordo, New Mexico, when I had to testify in two separate court-martials in the same week. The missions were accomplished.

Photo: Courtesy of U.S. Air Force

I was being presented the Air Force Commendation Medal for meritorious service by Brigadier General Beyea, Commander AFOSI Headquarters, when he was visiting Kunsan Airbase, Korea, in December 1982. One of my contributions was the construction of a combat ground-intelligence network, which consistently surpassed all expectations.

Photo: Courtesy of U.S. Air Force

I was being presented an Air Force Commendation Medal from Colonel Glenn Cox, Commander AFOSI District 10. Mine was the fifth-such medal awarded during Colonel Cox's six-year tenure as District 10 commander. Colonel Cox was ahead of his time when he employed the management-by-objectives program to track operational progress within his command. By using statistics to track each objective, he was able to determine if each detachment on a military base was meeting expectations or not. After we both retired from the Air Force, I continued to associate with Colonel Cox until his passing in the early 1990s.

Photo: Courtesy of U.S. Air Force

This photograph is of Brigadier General Brooksher, Commander Air Force Security Forces, and me. Brigadier General Brooksher was at Laughlin AFB, Texas, to congratulate their Security Forces squadron as the "best" small squadron in the Air Force. Brigadier General Brooksher stopped to visit our two OSI agents. The general always enjoyed a funny story; here I am telling him a "whopper."

Photo: Courtesy U.S. Air Force

This picture from 1986 at the San Antonio Logistic Center is representative of large OSI detachments assigned to locations where billions of dollars of aircraft parts are maintained for Air Force installations worldwide. I am on the far right, and Lieutenant Colonel Reid, wearing the Class A uniform, was commander of the detachment. Lieutenant Colonel Reid was highly skilled in OSI operations and management.

GUARDIANS OF AIR FORCE RESOURCES

Photo: Courtesy U.S. Air Force

Close friends—This picture, which was taken in 1981, is of the OSI team in Kunsan, Korea. These team members were in their twenties; I (front left) was in my forties and fondly addressed as "Dad." Brigadier Beyea, Commander AFOSI Headquarters, informed us that Jerry (front right) was selected to attend the Air Force Commissioning Program and polygrapher school (to administer polygraph examinations). When I was reassigned to San Antonio, Texas, Jerry would follow to attend the commissioning program. (He loved Mexican food!) However, when Jerry got to the US, he was diagnosed with cancer and passed away at age twenty-seven—a big blow to the rest of the team.

In 1982, when I transferred from Kunsan Air Base, South Korea, I was presented the plaque above. The presentation took place in the "war room" of the Kunsan City Police Headquarters. The senior superintendent (chief of police) made the presentation. The plaque reads as follows:

> This represents the deep appreciation for your immeasurable efforts in promoting the relationship between Korea and America during your tenure, June 1981–June 1982, as a Special Agent of the AFOSI. We greatly appreciate your untiring assistance rendered regarding matters of regional security and law enforcement conducted by the Korean National Police and AFOSI. We wish you and your family good fortune. May you always be successful in your future endeavors.

The award above is presented to agents who retire from the OSI and the United States Air Force. It includes the OSI badge with the agent's initials and is encased in acrylic. The brass plate shows the agent's name, rank, and years in the OSI.

When I retired from the OSI in 1987, I was surprised by the presentation of this plaque by the FBI office in San Antonio, Texas. I will always cherish the plaque, especially the FBI seal. The FBI agents were so kind to sign their names on its reverse side.

MY AMUSING TIDBITS

1. This happened to OSI agents conducting background checks: An elderly lady answers the knock at the door, looks at the agents' credentials, and says, "Nope. I've never seen him."

2. I came to a vehicular gate; the ranch house was about thirty yards away. I opened the gate and decided to walk to the house. On the way, I came face-to-face with a humongous Brahma bull. As the bull charged, I ran towards the gate and threw my notepad over the gate as I used my hands to leap over it. To this day, I do not know how I did it.

3. In a nice neighborhood, I knocked on a door, and no one answered. The house had a large wraparound porch. I walked around the porch and came face-to-face with a huge dog (a Saint Bernard). This time, I stood still—friendliest dog I've ever met.

4. Unlike the Saint Bernard incident, this time it was a Chihuahua. I knocked on a door, and this lady, after I identified myself, invited me in. Just then, her phone rang somewhere else in the house. She excused herself and said she would be right back. Then, there came Fifi, the Chihuahua, standing in front of me, shaking, and growling. Suddenly, she bit my pantleg (my new pants). I swung my leg; Fifi slid on the shiny wooden floor and ran off. When the lady returned, she asked, "Did Fifi keep you company while I was gone?"

5. Once, I tagged along with a police lieutenant and a Texas ranger (names of *John* and *Joe* are not real names), both in their sixties, while they were executing a search warrant. The suspect and his wife were military members. At the residence, the three of us went into separate rooms. The Texas ranger noted an object on the floor

next to the bed. He called out to the lieutenant and asked, "What is that?" The lieutenant replied, "That is one of those smoking pipes from Thailand." I walked in and was asked by Joe what the object by the bed was, and I replied, "That is a dildo used in sex variety." Joe looked at John and said, "We've been doing this too long." Joe said to me, "You say anything to anyone, I will hurt you!"

6. One time the base commander called me and said, "Grab your golf gear and meet me at the golf course." I had played with him before, so I did not think the request was odd. When I arrived at the golf course, I noted a military vehicle with four stars on the bumper's license plate. The base commander introduced me to the four-star general and the general's aide, a major. We started to play after agreeing on a small wager. On the seventeenth hole, the general and his aide won the hole and tied the base commander and me. On the eighteenth and last hole, there was a lake to the right and a public road on the left (out of bounds, OB). The general was up and pushed his shot in the lake. The major pulled his shot to the left, over the road (OB). The base commander also pushed his shot into the lake. I was last to hit. The base commander walked over to me and whispered, "I don't want to lose to this general on my own course, so keep the ball on the fairway and THAT IS AN ORDER." I had never heard of such a military order, but I did not want to lose either. I hit the ball down the fairway; the base commander and I won the match.

7. Speaking about generals, I was part of a protective detail on the ground when President Reagan in Air Force One was approaching to land on an Air Force base. I was posted inside Base Operations, close to where the commanding general (12 stars) and the local city mayor were waiting for the president of the United States. I overheard a woman's voice say, "*Jonathan* (not real name of the general), stop pacing up and down. You are making me nervous." Everyone has a boss.

8. Another protective detail I was involved in took place in New York City during the United States General Assembly Conference. The detail involved the protection of Prime Minister Fidel Castro.

During his stay in New York in the Cuban Mission Building, the New York Police Department (NYPD) established a cordon two blocks around the Cuban Mission Building. Residents within the cordon had to be escorted, after proper identification, to their residences by a member of the protective detail. Only four Americans were allowed to visit Prime Minister Castro. It was my turn to escort the fourth visitor. It was raining and very cool when I got to the police cordon. There was a petite lady (five feet tall) wearing a raincoat and using an umbrella. As we walked to the Cuban Mission Building, I broke the silence by saying, "You look very familiar." She replied, "Of course, I'm Barbara Walters." Enough said.

9.   During another Protective Services operation in Las Vegas, Nevada, another agent and I operated the command post (CP). The principal was the air marshal from Great Britain. When the principal arrived at the airport, he was picked up by the motorcade detail, which transported the principal and his wife to the hotel. As things were getting ready for the principal's arrival, the Las Vegas metro police and their bomb-detecting dog examined the hotel suite where the principal and his wife were going to be staying. The bomb dog suddenly alerted us by barking at the ceiling. In the meantime, the principal's motorcade was one block from arrival. I made the decision for the motorcade not to stop at the hotel while the metro police were looking in the room's ceiling for what the bomb dog detected. The motorcade leader informed the CP that the principal had to use the restroom as soon as possible. Again, the CP advised the motorcade not to stop at the hotel. Finally, the metro police ascertained that the bomb dog had alerted on the air filter in the ceiling. The motorcade leader radioed in and said it was "very urgent" the motorcade stop for the principal to use the restroom. I informed the motorcade leader to go ahead and arrive at the hotel. Everything was okay, and the restroom in the lobby was secured with plainclothes police officers standing by. The air marshal, with agents along, practically ran into the lobby. He was happy again.

10. When I was stationed in South Korea during 1981, I became close friends with a captain in the Korean National Police. I spoke no Korean, and he spoke no English, but I always had our interpreter with me when I went to see him. One day on the way to see the police captain, Mr. Yi (the interpreter) and I walked by a huge sports building with many Ping-Pong tables. After our business, I asked the captain if he played Ping-Pong, and he said yes. I had played the game over the years, and when I was in junior high school, I was the Ping-Pong champion of my school (seventh grade in 1954). We walked over to the gym, and I noticed the captain was carrying a small case. At the gym, the captain took out a paddle from the case, removed his coat, and rolled up his sleeves. When we started playing, I immediately had to stand between five and ten feet away from the table to return the ball every time the captain slammed it. Then, there was the crowd who had stopped playing to watch. It is only logical that the player who stands away from the table to play defense usually loses to the player who, playing offense, stands in front of the table. I lost and commended the captain on his excellent play. We shook hands and had a good laugh.

LIFE AFTER OSI

Photo: Courtesy U.S. Army

After OSI retirement in 1987, I went to work for the U.S. Army at Fort Sam Houston in San Antonio, Texas, as a physical securtiy inspector. I was a member of a four-man team under the Provost Marshal's Office (PMO), which could inspect any of the U.S.

Army facilities to ensure they would meet the minimum regulatory physical-security standards.

About three years later, I was selected to fill the deputy provost marshal position (as a civilian). The provost marshal (PM) was a member of the garrison staff. I was responsible for the day-to-day operations of a company of one hundred military police (MP) soldiers, including a company commander and first sergent. I also maintained oversight of the one hundred contract guards we employed; they operated the eleven access control points (ACPs), also known as gates. During the deployment of 98 percent of the MP soldiers to Afghanistan in 2008, I was tasked to convert one hundred MP soldiers and one hundred contract guard positions to government civilian police officers. This conversion occurred at several U.S. Army installations. After the conversion I ended up with 200 civilian police officers.

The Department of the Army and Law Enforcement Operations later changed my title to chief of police. Not long after, it was time to retire at age seventy-two.

Photo: Courtesy U.S. Army

I had a police force of 200, and the officers above were my specialized folks. The military calls them a special reaction team (SRT). The civil-law police agencies call them a special weapons and tactics (SWAT) team. I ensured my team received all the proper training, as well as training alongside local SWAT agencies. The training prepared the SRT for such emergencies as active-shooter and hostage situations. In January 2006, then-President Bush visited the wounded soldiers at FSH and Brook Army Medical Center (BAMC). The SRT supplemented the Secret Service in covering the president when he walked from one building to the next. The photo above was taken at BAMC after their assignments.

ABOUT THE BOOK

GUARDIANS OF AIR *Force Resources* contains several condensed versions of investigated criminal cases with sufficient detail for easy understanding. Each chapter represents a summary of a real-life criminal case. *Guardians of Air Force Resources* also contains a section addressing amusing "tidbits" that were encountered by author Ray Acuña. Another section includes pictures of Ray Acuña's memorable moments as a U.S. Air Force Office of Special Investigations (AFOSI) agent. The last section of the book describes what the author did after his retirement from AFOSI; he talks of his work as a civilian employee working with the U.S. Army in law enforcement.

ABOUT THE AUTHOR

MR. ACUÑA ENLISTED in the U.S. Air Force in 1961. In 1965, Mr. Acuña served two combat tours in Vietnam. While there, he had the distinct pleasure of meeting the first US president ever to visit a combat zone: Lyndon B. Johnson, who in conversation referred to himself and Mr. Acuña as "fellow Texans." After returning to the United States, Mr. Acuña continued his career in the Air Force and obtained an associate of arts degree in criminal justice.

In 1973, he was selected to the U.S. Air Force Office of Special Investigations (AFOSI) as a special agent, with subsequent attendance of the OSI Academy. He served in that capacity until retirement in 1987. In 1989, Mr. Acuña went to work in an Army installation in the field of law enforcement, serving as the deputy provost marshal, a position he held until retirement in 2011.

After fifty years of government service, Mr. Acuña currently resides with his wife, Roselia, in the state of Texas.

www.ingramcontent.com/pod-product-compliance
Lightning Source LLC
LaVergne TN
LVHW020431080526
838202LV00055B/5122